At Sylvan, we believe that a lifelong love of learning begins at an early age, and we are glad you have chosen our resources to help your children experience the joy of mathematics as they build critical reasoning skills. We know that the time you spend with your children reinforcing the lessons learned in school will contribute to their love of learning.

Success in math requires more than just memorizing basic facts and algorithms; it also requires children to make sense of size, shape, and numbers as they appear in the world. Children who can connect their understanding of math to the world around them will be ready for the challenges of mathematics as they advance to more complex topics.

We use a research-based, step-by-step process in teaching math at Sylvan that includes thought-provoking math problems and activities. As students increase their success as problem solvers, they become more confident. With increasing confidence, students build even more success. The design of the Sylvan workbooks will help you to help your children build the skills and confidence that will contribute to success in school.

Included with your purchase of this workbook is a coupon for a discount at a participating Sylvan center. We hope you will use this coupon to further your children's academic journeys. Let us partner with you to support the development of confident, well-prepared, independent learners.

The Sylvan Team

Sylvan Learning Center
Unleash your child's potential here

No matter how big or small the academic challenge, every child has the ability to learn. But sometimes children need help making it happen. Sylvan believes every child has the potential to do great things. And, we know better than anyone else how to tap into that academic potential so that a child's future really is full of possibilities. Sylvan Learning Center is the place where your child can build and master the learning skills needed to succeed and unlock the potential you know is there.

The proven, personalized approach of our in-center programs deliver unparalleled results that other supplemental education services simply can't match. Your child's achievements will be seen not only in test scores and report cards but outside the classroom as well. And when he starts achieving his full potential, everyone will know it. You will see a new level of confidence come through in everything he does and every interaction he has.

How can Sylvan's personalized in-center approach help your child unleash his potential?

- Starting with our exclusive Sylvan Skills Assessment®, we pinpoint your child's exact academic needs.

- Then we develop a customized learning plan designed to achieve your child's academic goals.

- Through our method of skill mastery, your child will not only learn and master every skill in his personalized plan, he will be truly motivated and inspired to achieve his full potential.

To get started, included with this Sylvan product purchase is $10 off our exclusive Sylvan Skills Assessment®. Simply use this coupon and contact your local Sylvan Learning Center to set up your appointment.

And to learn more about Sylvan and our innovative in-center programs, call 1-800-EDUCATE or visit www.SylvanLearning.com. *With over 1,000 locations in North America, there is a Sylvan Learning Center near you!*

Kindergarten
Math
Games & Puzzles

Copyright © 2010 by Sylvan Learning, Inc.

Published in the United States by Random House, Inc., New York, and in Canada by Random House of Canada Limited, Toronto.

www.tutoring.sylvanlearning.com

Created by Smarterville Productions LLC
Producer & Editorial Direction: The Linguistic Edge
Producer: TJ Trochlil McGreevy
Writer: Amy Kraft
Cover and Interior Illustrations: Shawn Finley and Duendes del Sur
Layout and Art Direction: SunDried Penguin
Director of Product Development: Russell Ginns

First Edition

ISBN: 978-0-375-43033-6

This book is available at special discounts for bulk purchases for sales promotions or premiums. For more information, write to Special Markets/Premium Sales, 1745 Broadway, MD 6-2, New York, New York 10019 or e-mail specialmarkets@randomhouse.com.

PRINTED IN CHINA

10 9 8 7 6 5 4 3 2

Contents

Counting to 5

Connect the Dots

DRAW a line to connect the numbers in order, starting with 1.

Happy Clowns

DRAW balloons in each clown's hand to match the number the clown is holding.

Counting to 5

Mystery Picture

COUNT the dots, and COLOR each section according to the numbers to reveal the mystery picture.

1 =   2 =   3 =   4 =   5 =

Make a Match

CUT OUT the numbers and pictures. READ the rules. PLAY the game!
(Save these cards for use with page 11.)

Rules: Two players
1. Place the cards face down on a table.
2. Take turns turning over two cards at a time.
3. Keep the cards when you match a picture and a number.

The player with the most matches wins!

Lickin' Lollipops

Can you be the first to get to the delicious center of the lollipop? Use two small objects as playing pieces and the spinner from page 109. READ the rules. PLAY the game! (Save the spinner to use again later in the workbook.)

Rules: Two players
1. Place the playing pieces at Start.
2. Take turns spinning the spinner, and MOVE the same number of spaces.
3. Watch out for the sticky spaces. If you land on ✳ you lose a turn.

The first player to the delicious lollipop center wins!

Connect the Dots

DRAW a line to connect the numbers in order, starting with 1.

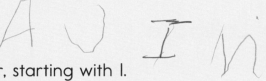

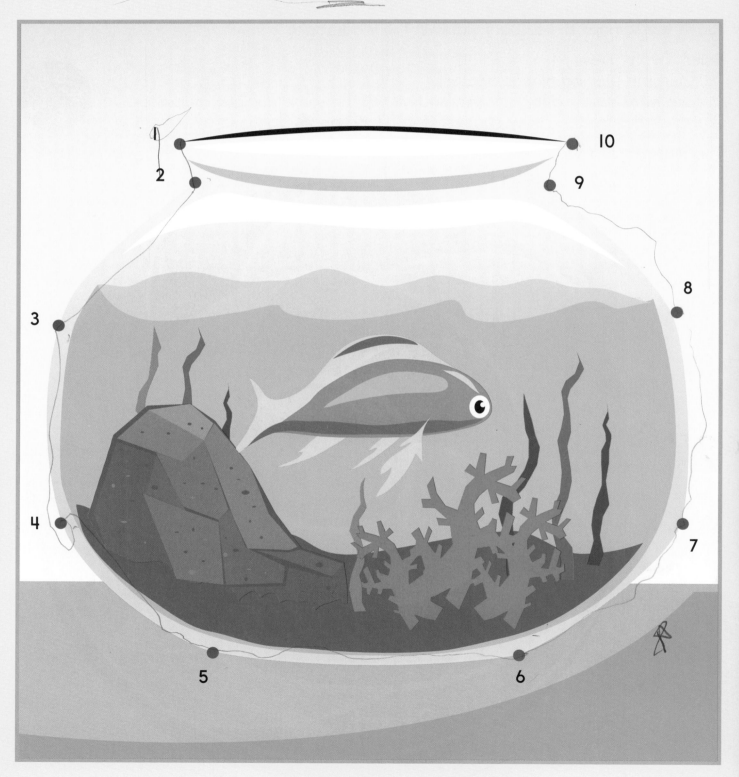

Happy Clowns

DRAW balloons in each clown's hand to match the number the clown is holding.

Mystery Picture

COLOR each section according to the numbers to reveal the mystery picture.

1 = 2 = 3 = 4 = 5 = 6 = 7 = 8 = 9 = 10 =

2

Make a Match

CUT OUT the numbers and pictures. READ the rules. PLAY the game!

HINT: Combine these cards with the cards from page 5 for a greater challenge.

Rules: Two players
1. Place the cards face down on a table.
2. Take turns turning over two cards at a time.
3. Keep the cards when you match a picture and a number.

The player with the most matches wins!

6

7

8

10 **9**

Hide and Seek

COUNT the number of times each animal appears in the picture. Then WRITE the number next to each animal.

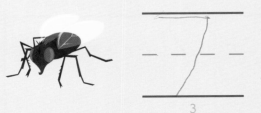
1

2

3

4

Domino Dots

Using the dominoes from pages 111 and 113, PLACE dominoes one at a time so that the touching sides of the dominoes have the same number of dots.
(Save the dominoes to use again later in the workbook.)

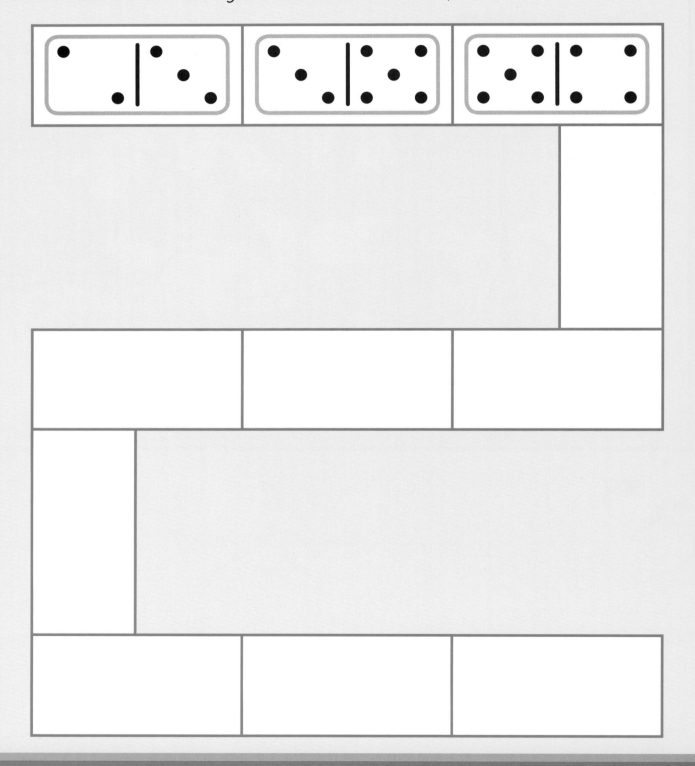

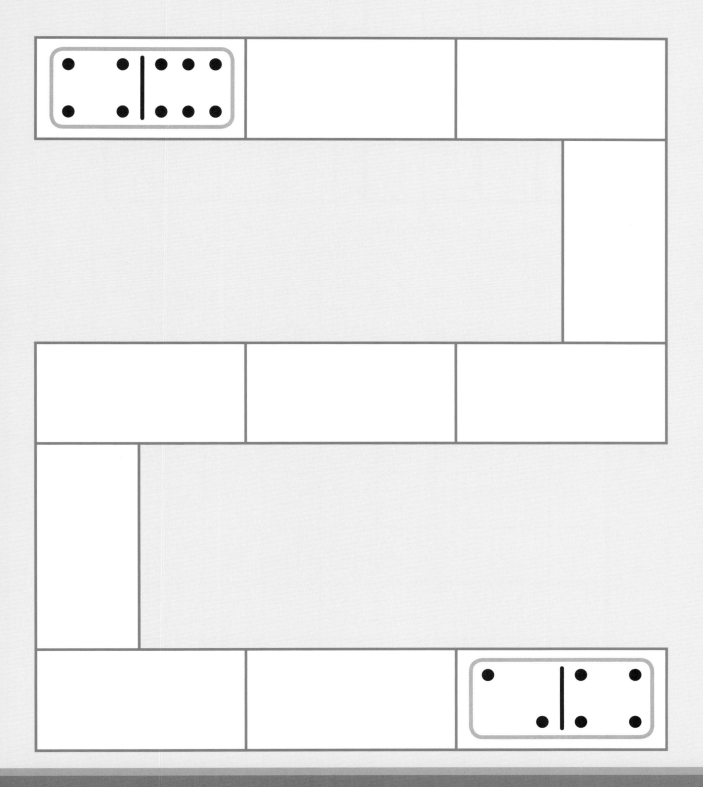

Super Sudoku

WRITE the numbers 1 through 4 so that each row, column, and box has all four numbers.

2	1	3	4
4	3	1	2
3	2	4	1
1	4	2	3

	2	1	
3			4
2	4		1
		4	

WRITE the numbers 1 through 9 so that each row, column, and box has all nine numbers.

3		8		4	7		5	2
	1	7	2		9		4	
9	2		5	8		3	7	1
4	9		7	1		2	8	
	8	5	3		4		9	7
6		1		2	8	5		4
7	3	6	4	9		8		5
	4		8			7	6	
8		2	6	7	3		1	9

Rat Race

Can you be the first rat to reach the cheese? Use two small objects as playing pieces and the spinner from page 110 (the backside of the spinner from page 109). READ the rules. PLAY the game!

Rules: Two players
1. Place the playing pieces at Start.
2. Take turns spinning the spinner, and move the same number of spaces.
3. If you land on a space with a number, move that many spaces in the direction of the arrow.

The first player to get to the cheese wins!

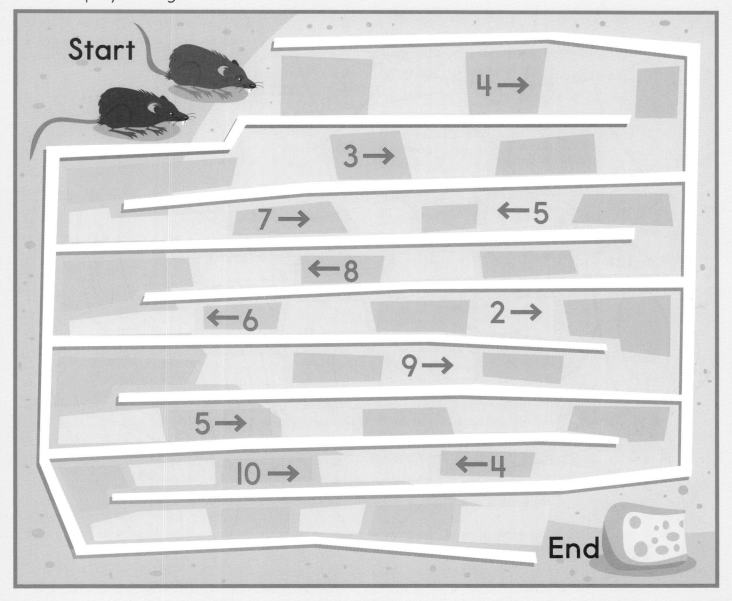

Mystery Picture

COUNT the dots, and COLOR each section according to the numbers to reveal the mystery picture.

4 dots = ⬛ More than = ⬛ Less than = ⬛
4 dots 4 dots

X Marks the Spot

DRAW an X on any group of bugs that has less than 5.

Domino Dots

Using the dominoes from pages 111 and 113, PLACE dominoes so that both parts of the domino have more dots than the one pictured. After you place all nine, see if you can keep finding more. (Save the dominoes to use again.)

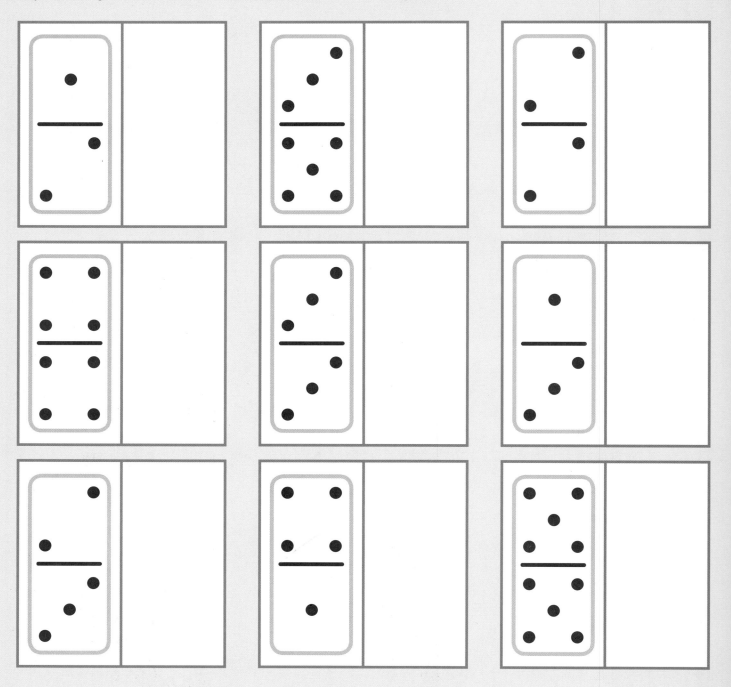

Your Deal

Using the number cards from a deck of playing cards, DEAL the cards so that the green spaces have one shape **less** than the card shown and the blue spaces have one **more** than the card shown.

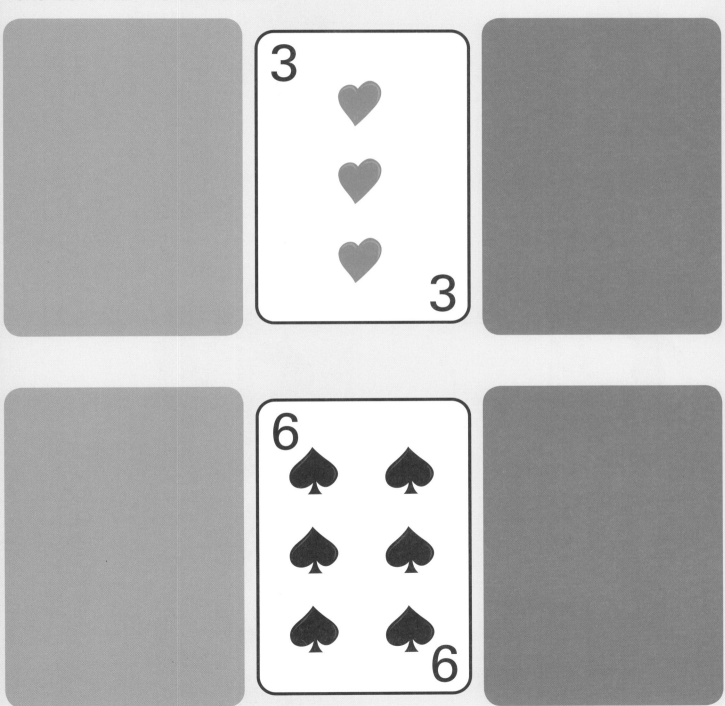

What's the Order?

WRITE the order of the pictures from 1st to 6th.

3

6

5

1

4

2

Secret Message

WRITE a letter using the order of the letters at the top.

C S O E H U I Y D A

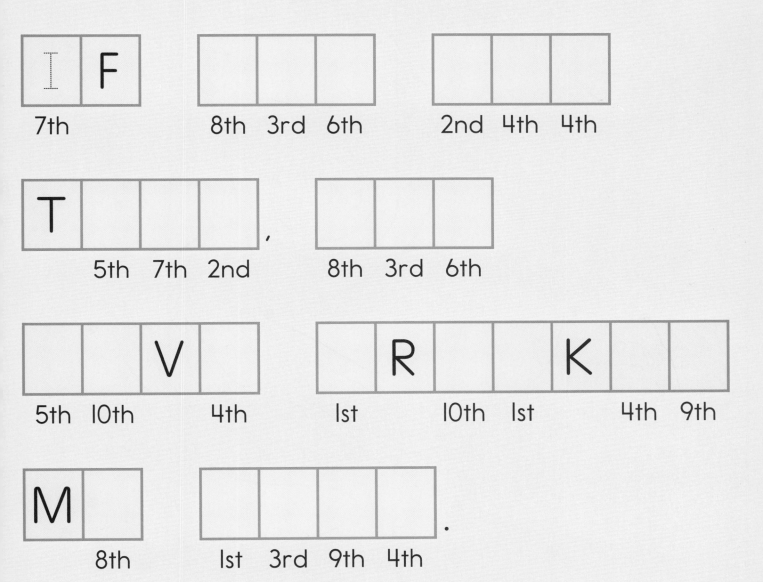

I	F
7th	

8th	3rd	6th

2nd	4th	4th

T		
5th	7th	2nd
,

8th	3rd	6th

		V
5th	10th	4th

	R		K	
1st	10th	1st	4th	9th

M	
8th	

1st	3rd	9th	4th
.

Connect the Dots

DRAW a line to connect numbers of the same color in order, starting with 1.

Domino Dots

Using the dominoes from pages 111 and 113, PLACE dominoes one at a time so that the touching sides of the dominoes have the same number of dots. (Save the dominoes to use again.)

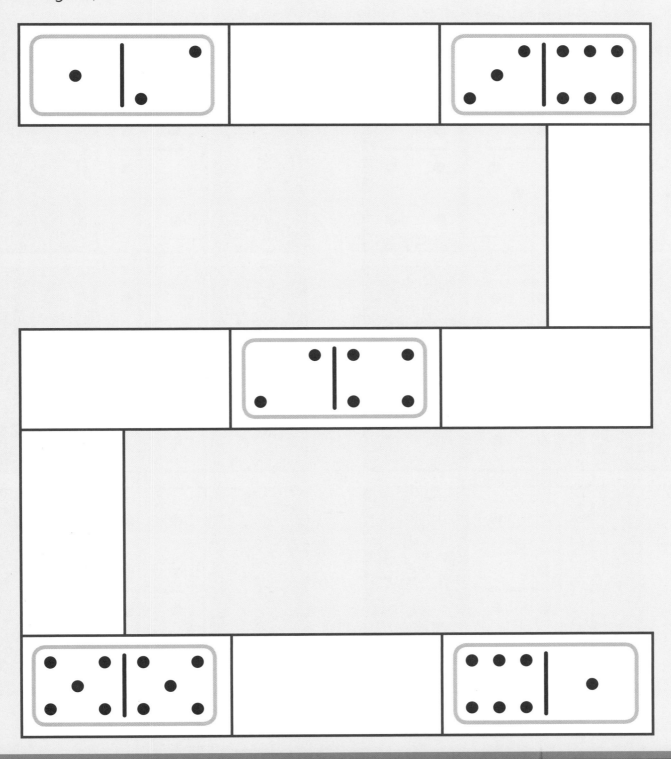

Domino Dots

Using the dominoes from pages 111 and 113, PLACE dominoes in the green spaces so that both parts of the domino have one dot **less** than the one pictured. PLACE dominoes in the blue spaces so that both parts of the domino have one dot **more** than the one pictured. (Save the dominoes to use again.)

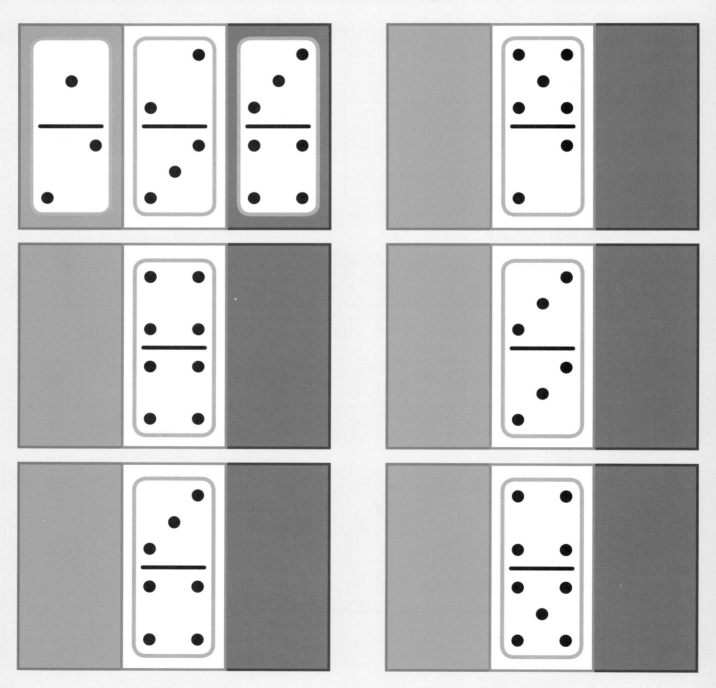

What's the Order?

WRITE the order of the pictures from 1st to 9th.

5

3

1

1

2

3

7

2

4

4

5

6

9

8

6

7

8

9

Spot the Differences

LOOK at the two pictures and CIRCLE the differences in the second picture.

HINT: There are eight differences.

Cross Out

DRAW an X on any box that does **not** exactly match the middle box.

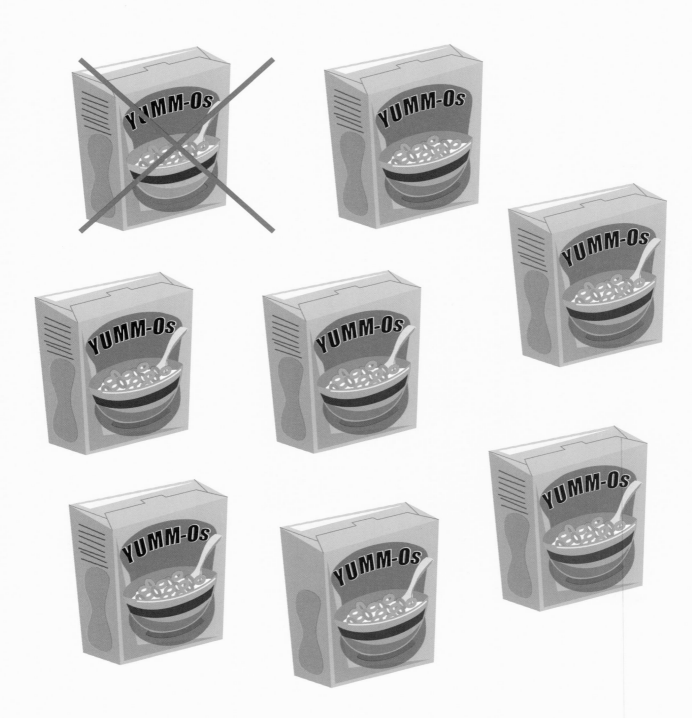

Make a Match

CUT OUT the pictures. READ the rules. PLAY the game!

Rules: Two players
1. Place the cards face down on a table.
2. Take turns turning over two cards at a time.
3. Keep the cards when you find two pictures that are the same.

The player with the most matches wins!

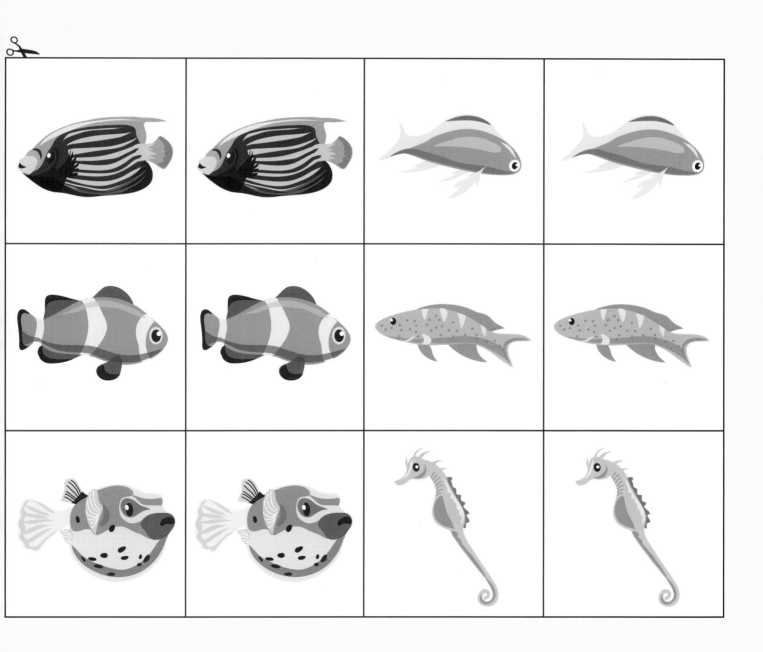

Two of a Kind

Only two of these action figures are identical. CIRCLE the matching pair.

Snake Spiral

FOLLOW the pattern and COLOR the snake.

Incredible Illusions

FOLLOW the pattern to finish coloring the picture. LOOK at the lines both before and after you color. Do they look straight?

HINT: Use a marker instead of a crayon.

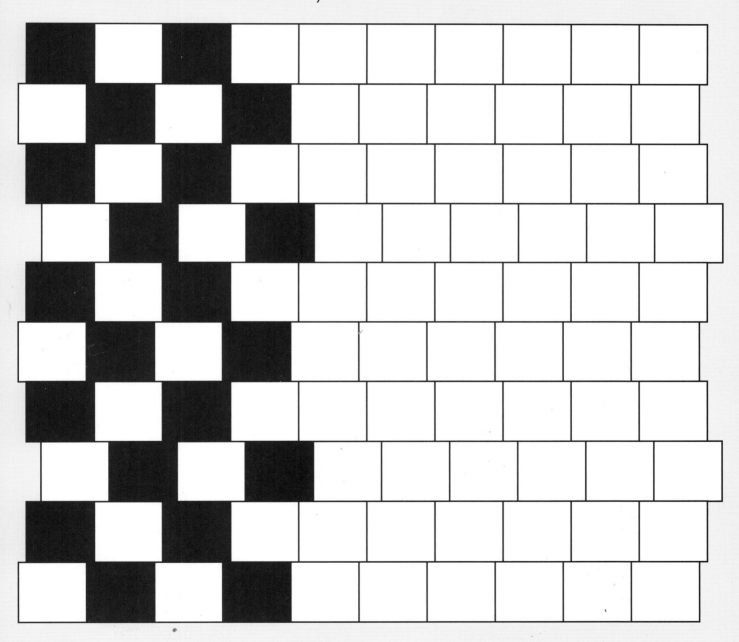

Picture Patterns

Pattern Maker

CUT OUT the shapes from the top of page 37, and PLACE them to finish each pattern.

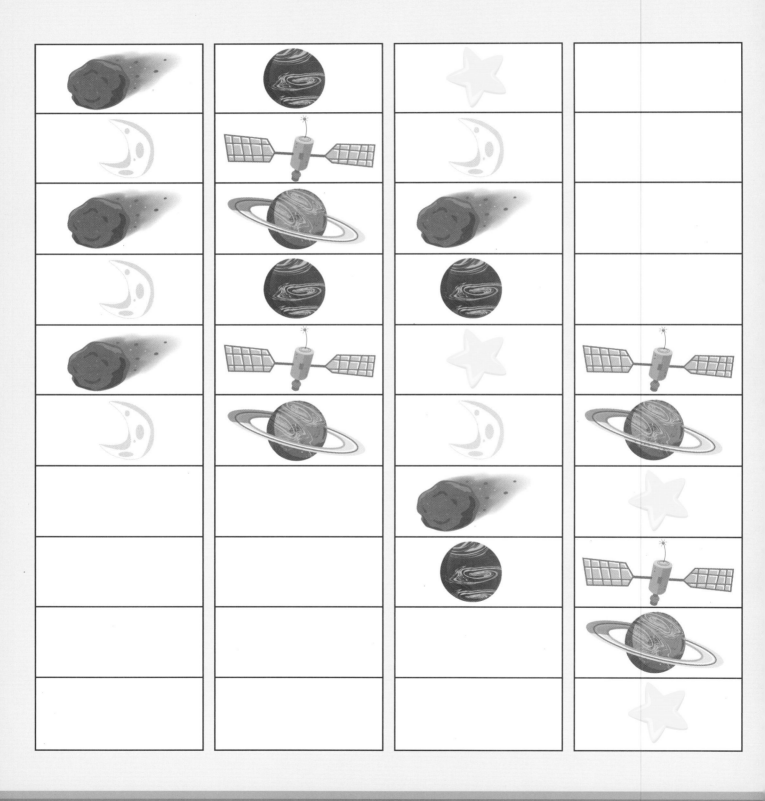

Use these shapes for page 36.

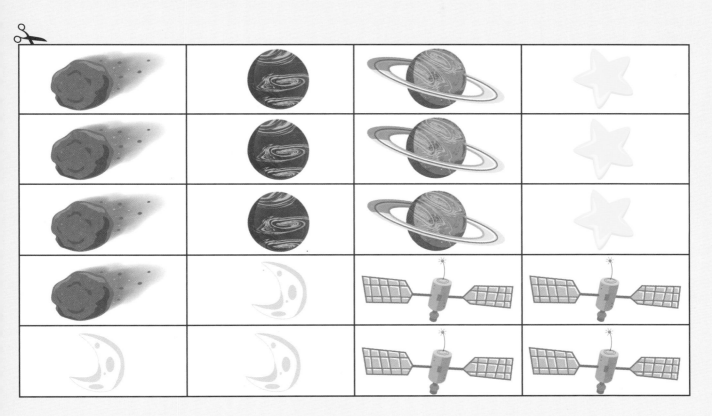

Use these shapes for page 39.

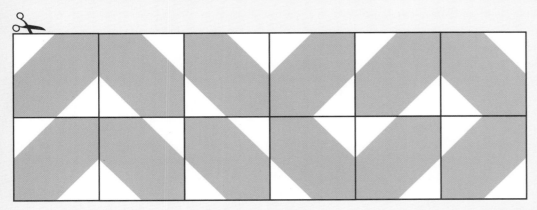

Pattern Maker

CUT OUT the shapes on the bottom of page 37, and PLACE them to finish each pattern.

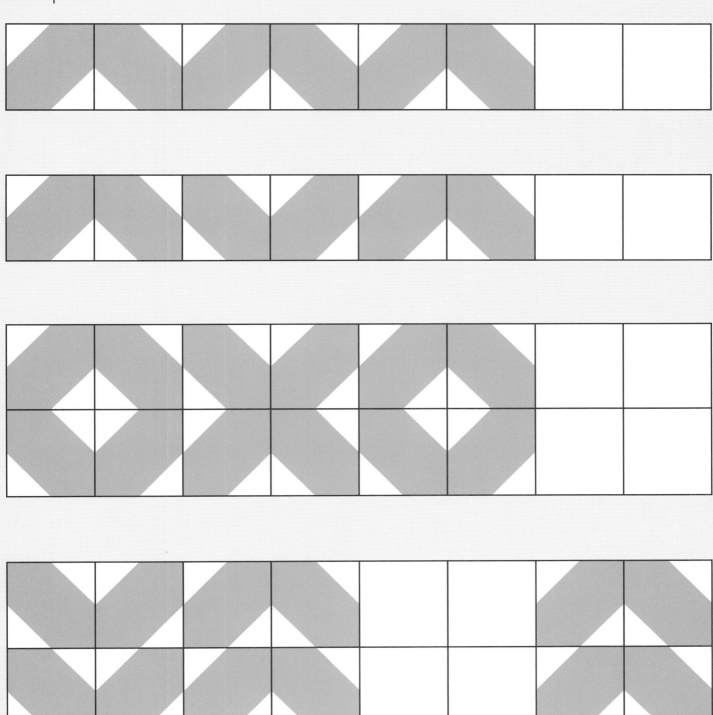

Get the Cheese

FOLLOW the path by connecting the numbers from 1 through 10 to get the cheese.

Dock the Boat

FOLLOW the path by connecting the numbers from 10 through 1 to dock the boat.

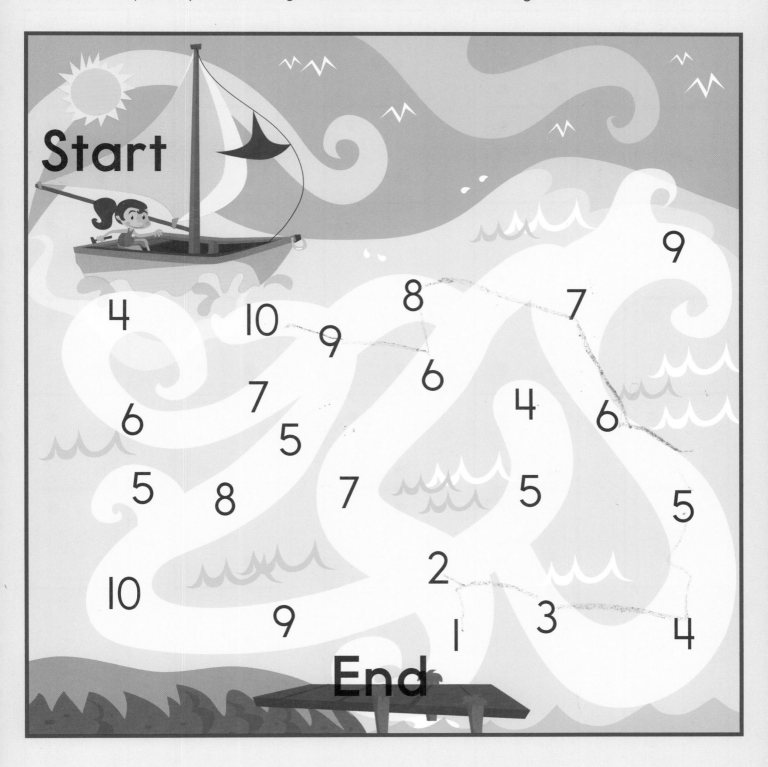

Domino Dots

Using the dominoes from pages 111 and 113, PLACE dominoes to finish each pattern.

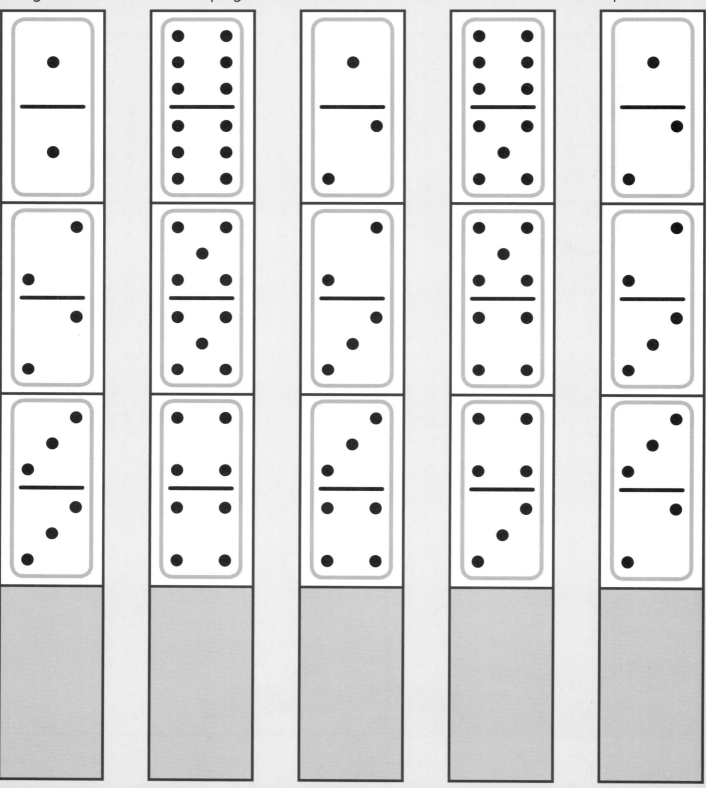

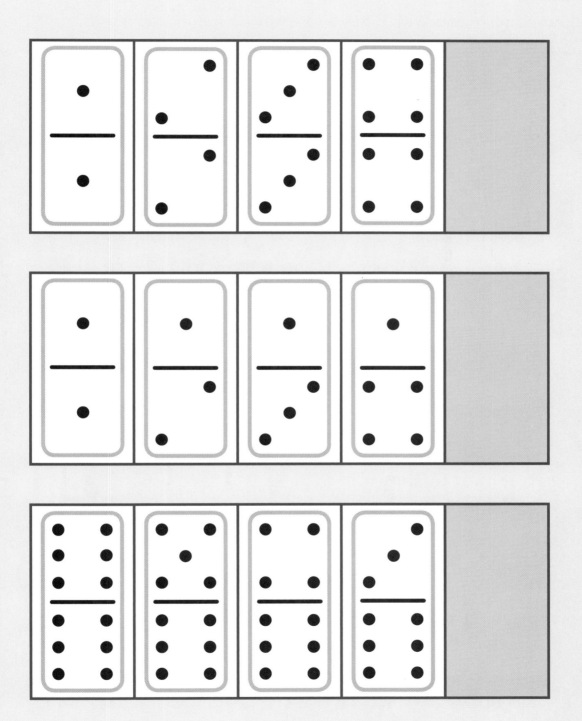

Number Patterns

What's Your Address?

FOLLOW the pattern, and WRITE the addresses on the mailboxes.

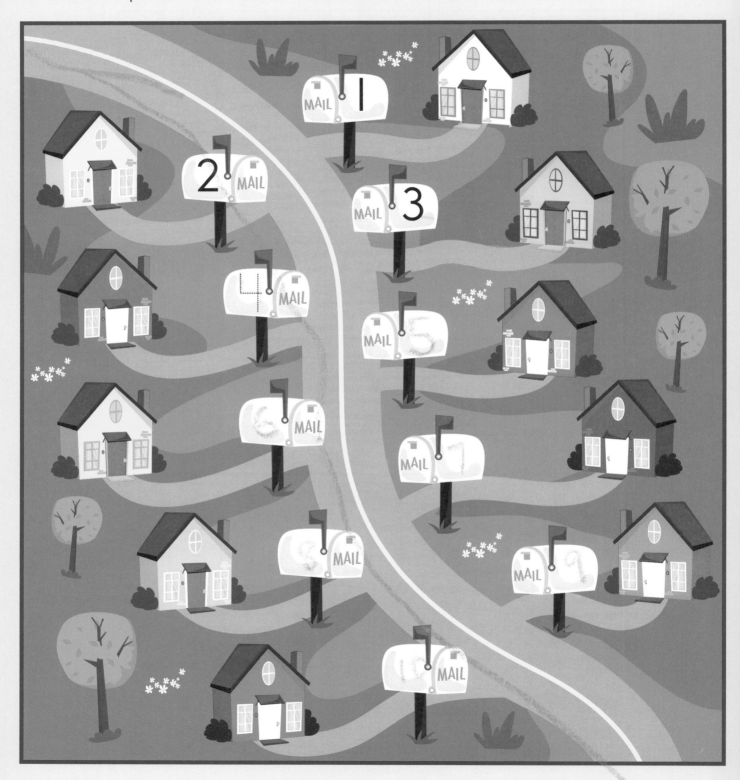

Your Deal

Using the number cards from a deck of playing cards, DEAL the cards shown in each row. DEAL two more cards in each row to finish each pattern.

1.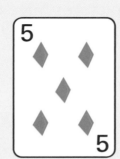

2.

3.

4.

Matched Set

CUT OUT the cards on page 47, and PLACE two cards next to each picture to make a matched set. (Save the cards to use on page 49.)

HINT: Look for things that the pictures have in common, like color, number, or anything else! See how many different sets you can make.

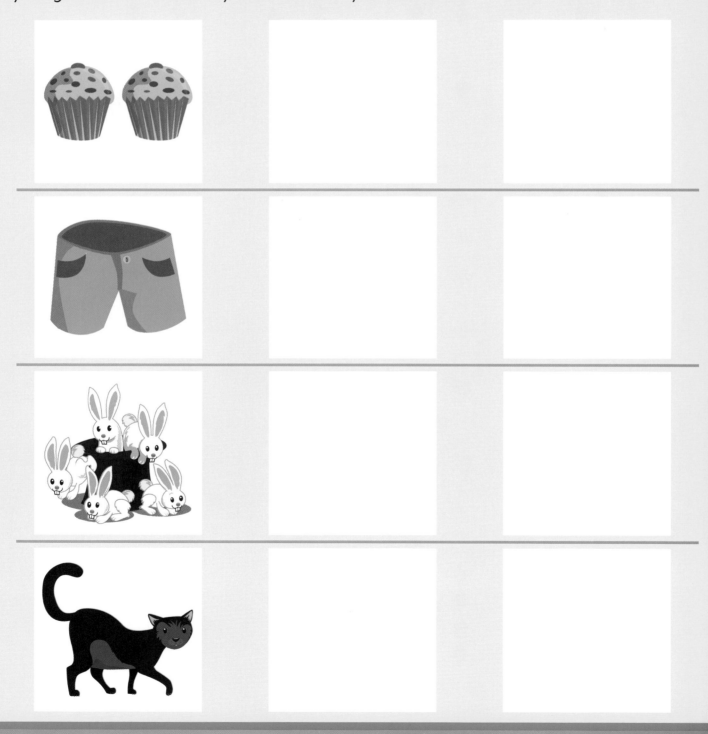

Matched Set

Using the cards from page 47, PLACE two cards next to each picture to make a matched set.

HINT: Look for things that the pictures have in common, like color, number, or anything else! See how many different sets you can make.

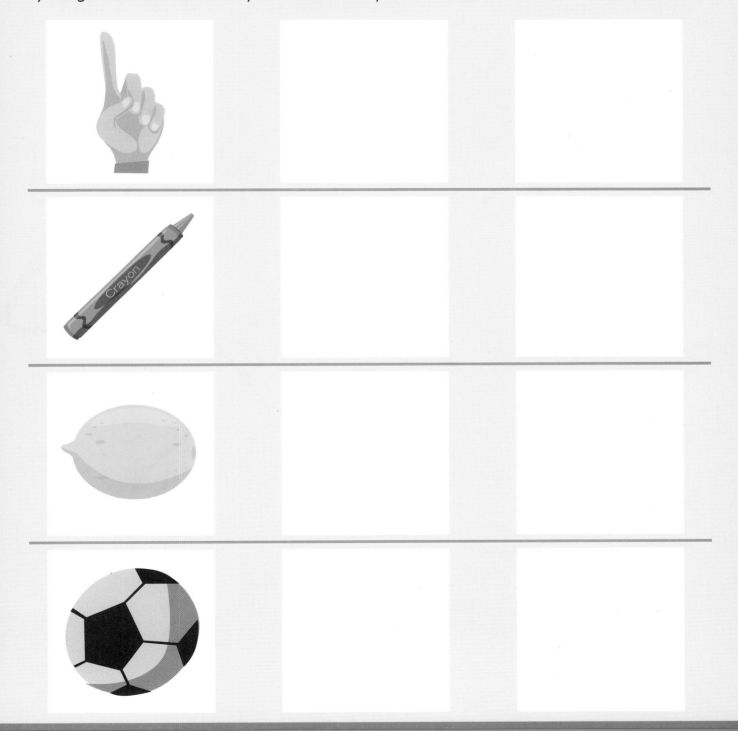

Picking Pairs

DRAW a line to connect each pair of objects that belong together.

Zookeeper

The new zoo signs are in the wrong places. DRAW a line from each sign to the place where it belongs.

Slide Sort

CIRCLE the objects that will fall in the wrong box when they go down the slide.

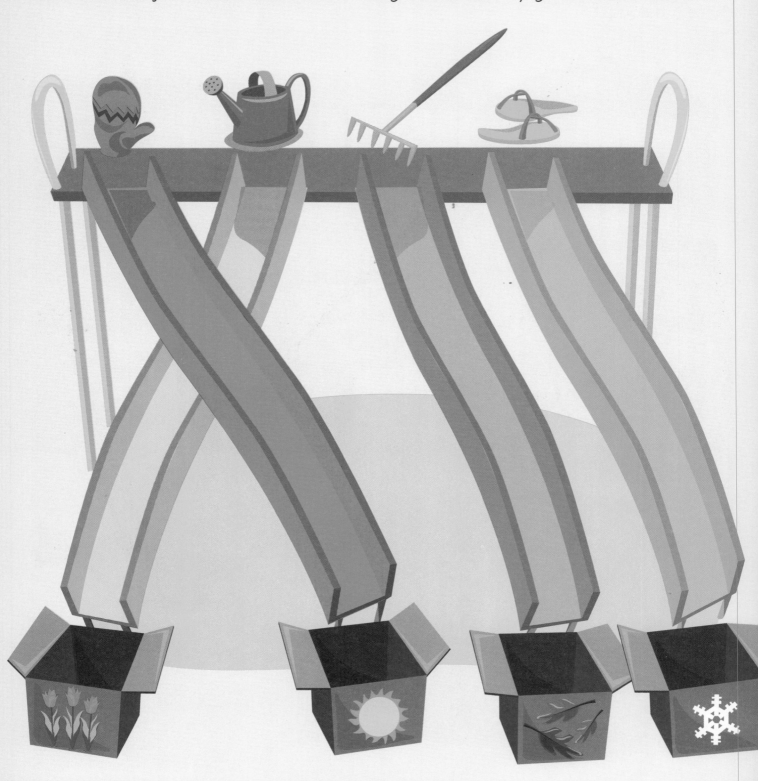

Picking Pairs

DRAW a line to connect each animal with its home.

Vexing Venn

CUT OUT the pictures on page 55, and PLACE them in the correct places in the Venn diagram.

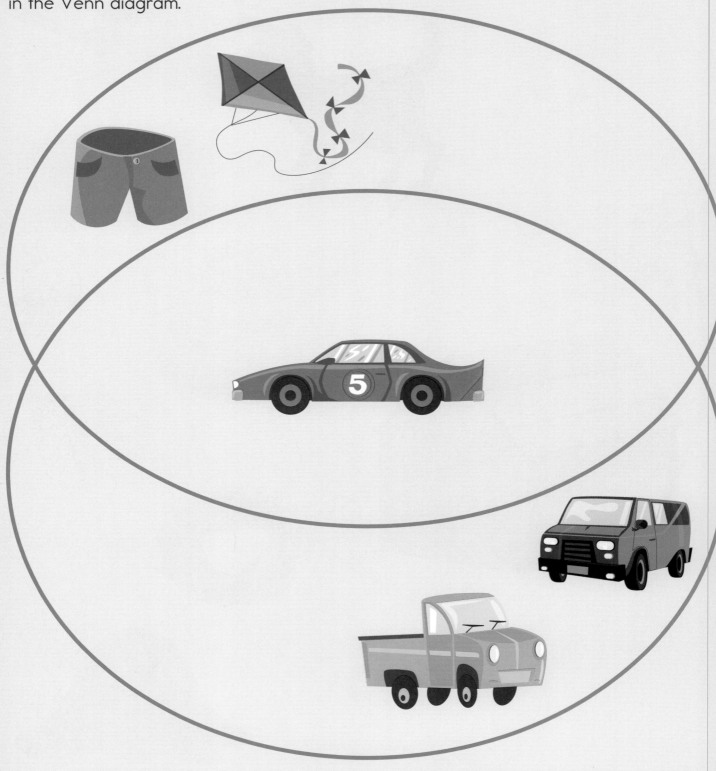

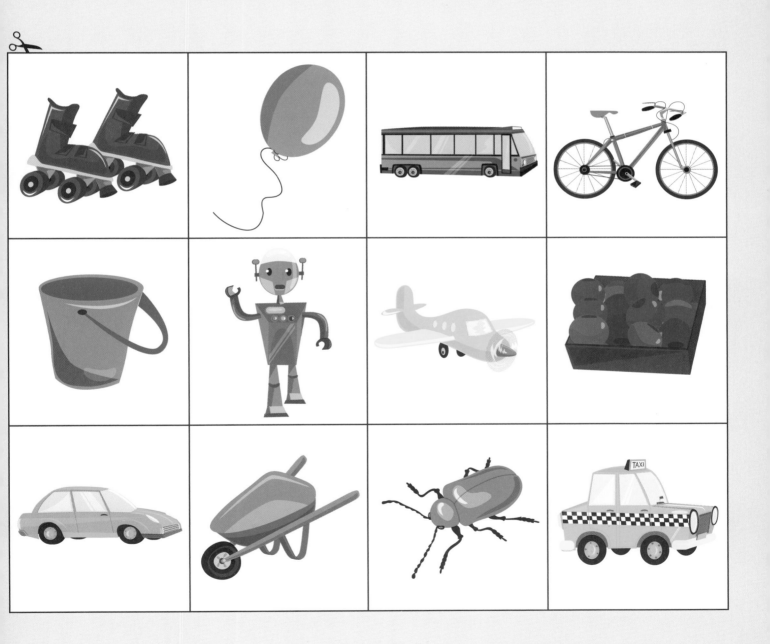

Pick for Packing

Help Nolan pack for a camping trip. DRAW lines from the objects he will need to his backpack.

Two of a Kind

Only two of these robots are identical. CIRCLE the matching pair.

Incredible Illusions

FOLLOW the pattern to finish coloring the picture using three colors. How many colors do you see when you've finished coloring?

HINT: Use markers instead of crayons.

Domino Dots

Using the dominoes from pages 111 and 113, PLACE dominoes to finish each pattern.

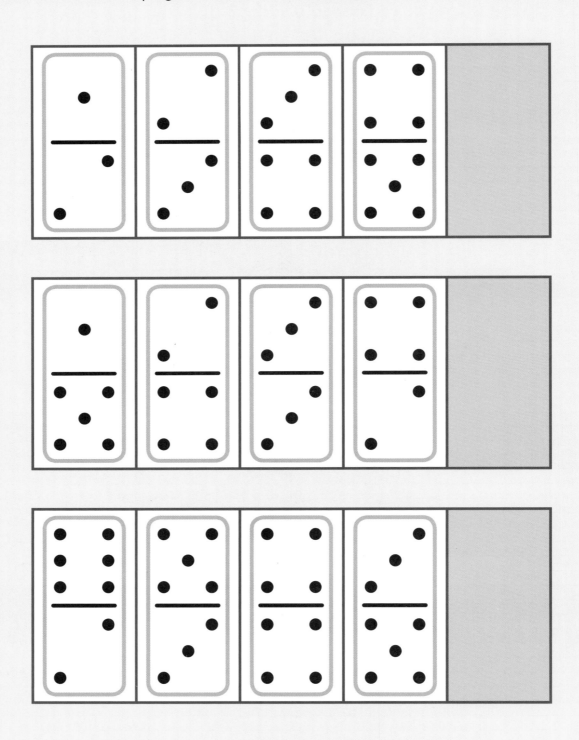

Picking Pairs

DRAW a line to connect each pair of objects that belong together.

Cross Out

DRAW an X on any shape that is **not** a circle.

circle

Cross Out

DRAW an X on any shape that is **not** a triangle.

triangle

Cross Out

DRAW an X on any shape that is **not** a square.

square

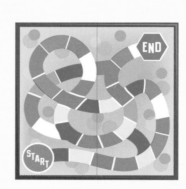

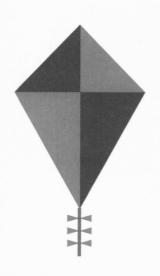

X Marks the Spot

DRAW an X on each rectangle hidden in the picture.

HINT: A square is a special type of rectangle.

Hidden Shapes

FIND each shape hidden in the picture. DRAW a line to connect each shape with its location in the picture.

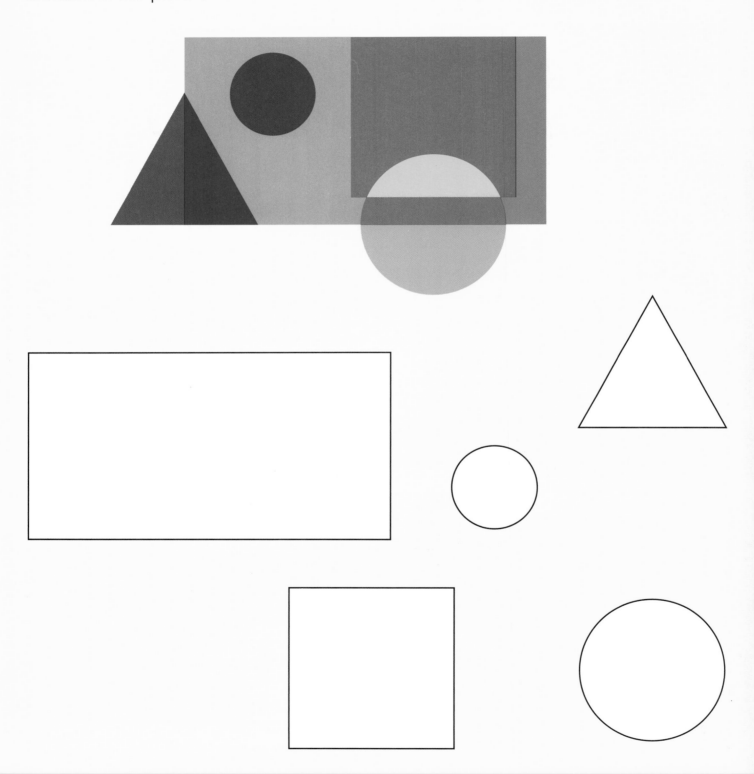

Make a Match

CUT OUT the pictures. READ the rules. PLAY the game!

Rules: Two players
1. Place the cards face down on a table.
2. Take turns turning over two cards at a time.
3. Keep the cards when you match two pictures.

The player with the most matches wins!

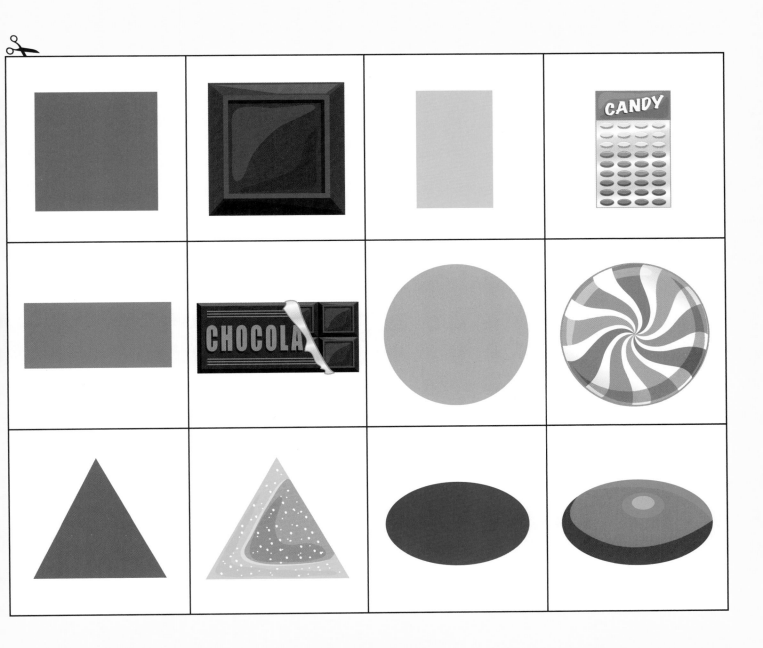

Hide and Seek

How many of each shape can you find in the picture? WRITE the number next to each shape.

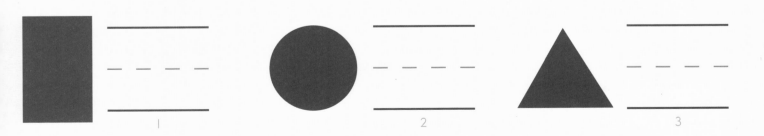

1

2

3

Shapely Sequence

DRAW and COLOR shapes to finish each pattern.

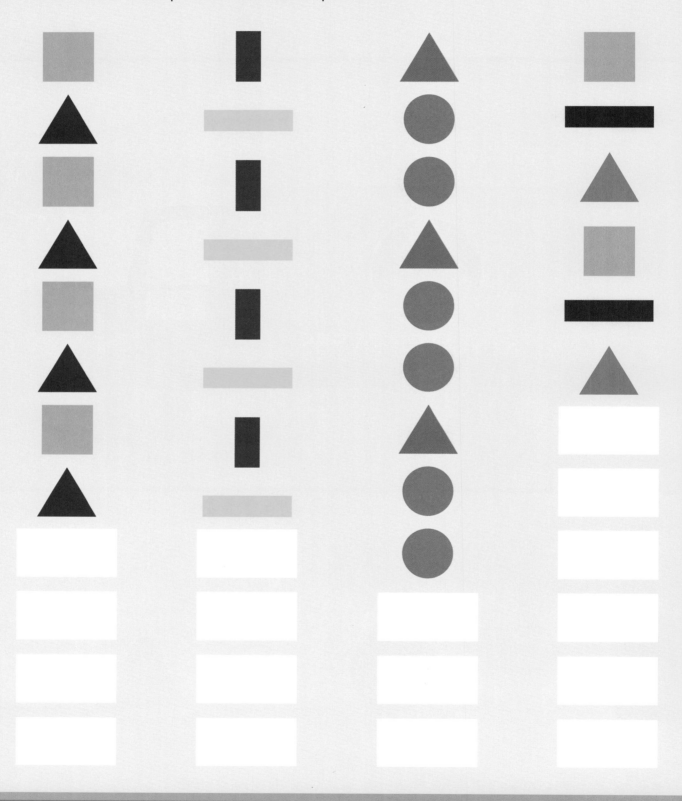

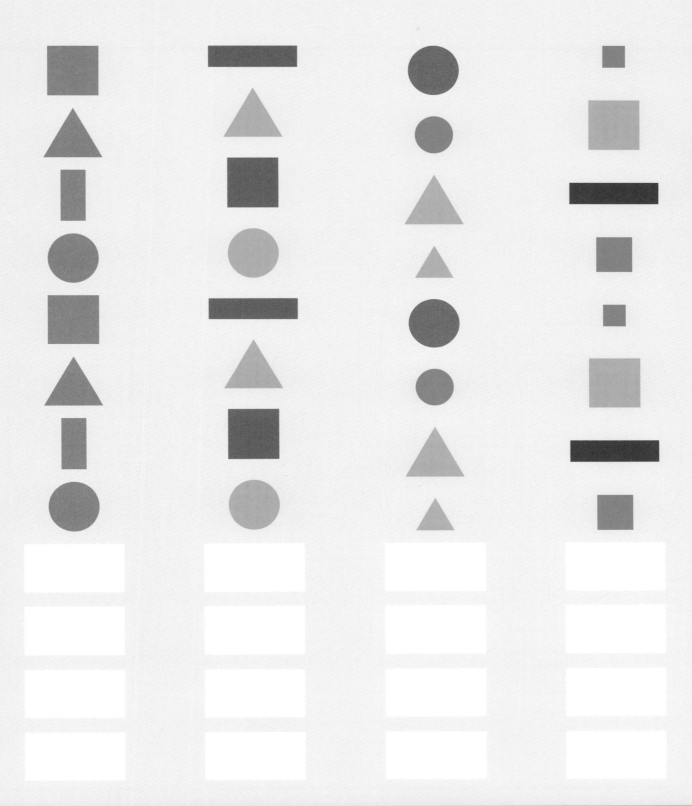

Picture Perfect

DRAW a circle on each robot's body to make a head. Then DRAW a face, and COLOR the robots.

Doodle Pad

TRACE the circles. Then DRAW a picture using each circle. Think about how many things are circles.

Picture Perfect

DRAW a triangle on each superhero's chest. Then DECORATE each triangle, and COLOR the superheroes.

HINT: You can make up names for the superheroes and put the first letter in the triangle, or draw their superpower.

Doodle Pad

TRACE the triangles. Then DRAW a picture using each triangle. Think about how many things are triangles.

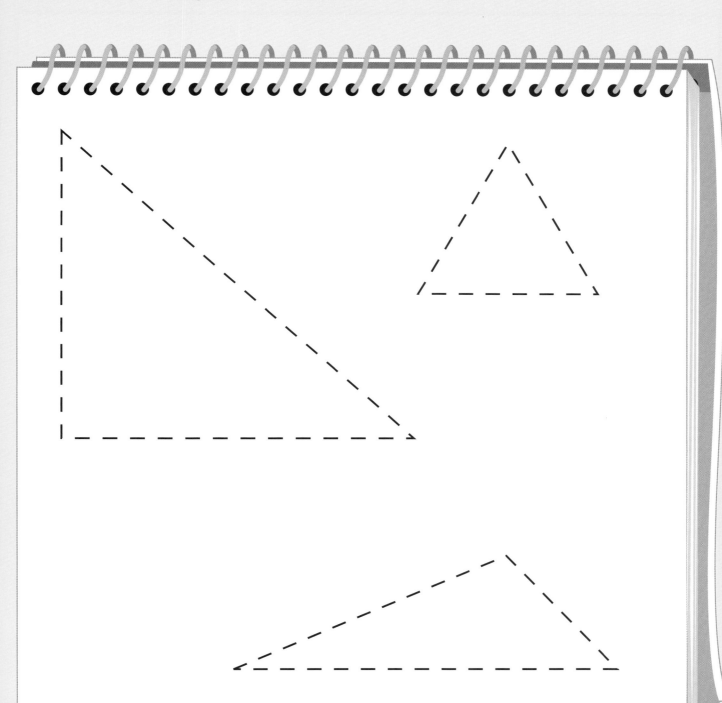

Picture Perfect

DRAW a rectangle around each doll to make a box. Then DECORATE the boxes and COLOR the dolls.

Doodle Pad

TRACE the squares. Then DRAW a picture using each square. Think about how many things are squares.

Tricky Tangrams

Use the tangram pieces from page 115, and PLACE the pieces to completely fill each shape without overlapping any pieces.

HINT: Try placing the biggest pieces first.

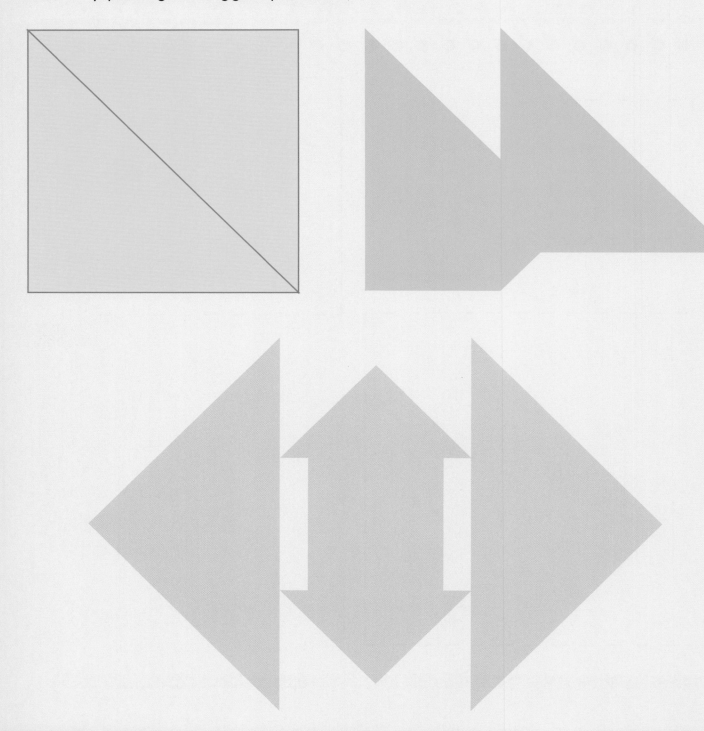

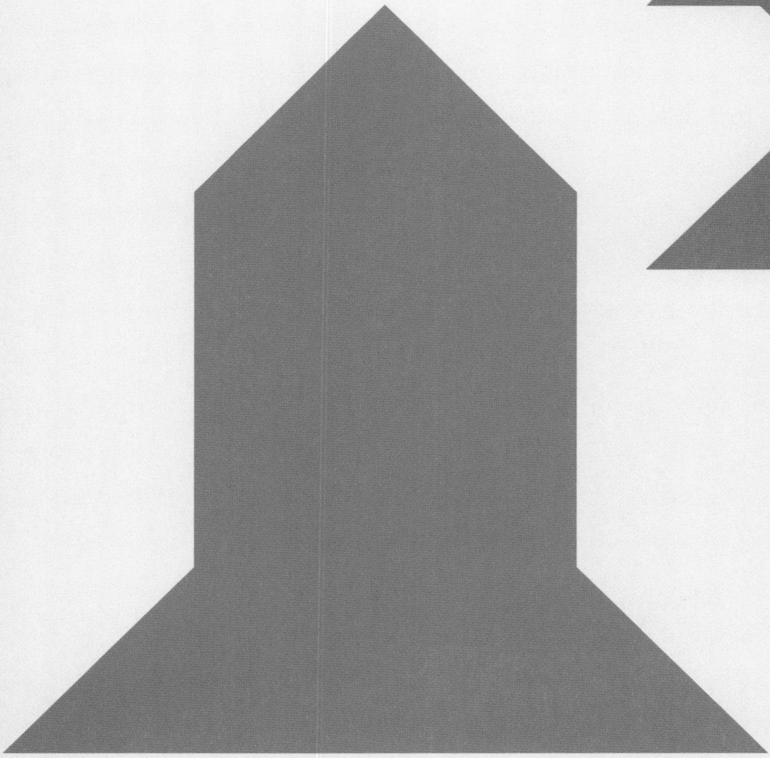

Shape Shifters

Use the pattern block pieces from page 117, and PLACE the pieces to completely fill each shape without overlapping any pieces. See if you can solve the puzzles different ways. (Save the pattern block pieces to use again.)

Shapes Squared

FOLLOW the directions in order, and DRAW and COLOR the shapes in the correct squares on page 83.

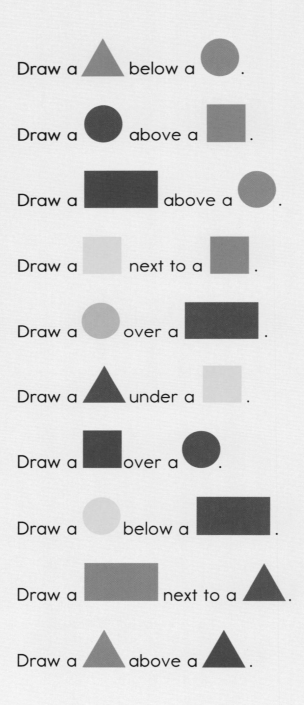

Draw a ▲ below a ⬤ .

Draw a ⬤ above a ◼ .

Draw a ▬ above a ⬤ .

Draw a ◼ next to a ◼ .

Draw a ⬤ over a ▬ .

Draw a ▲ under a ◼ .

Draw a ◼ over a ⬤ .

Draw a ⬤ below a ▬ .

Draw a ▬ next to a ▲ .

Draw a ▲ above a ▲ .

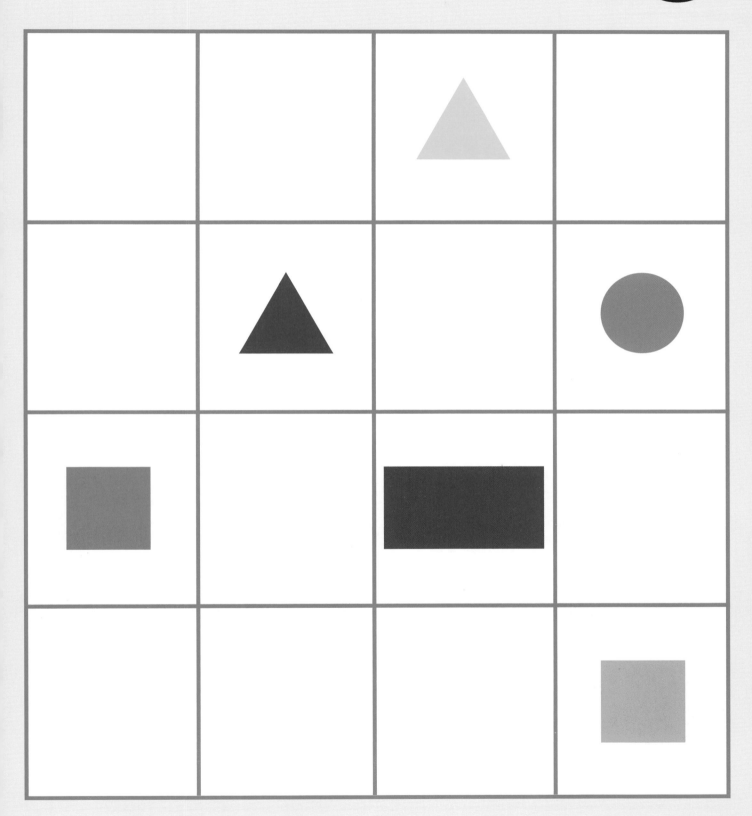

Magic Show

FOLLOW the directions to DRAW the parts of the magic show.

1. DRAW a rabbit under the table.

2. DRAW a rabbit inside the hat.

3. DRAW a magic wand over the hat.

Doodle Pad

DRAW three things above the chair and two things below the chair.

I'm Home!

DRAW a line from each person to the house where you think that person might live.

Treasure Hunt

There are many pirate treasures, but only one is real. FOLLOW the pirate's directions, and DRAW an X on the correct treasure.

Walk from my straight for a . Go around the past some .

Then head for the 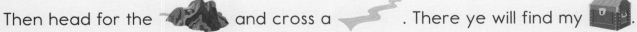 and cross a 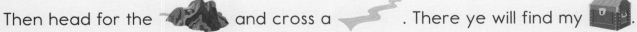. There ye will find my .

Where Am I?

FIND each person using the map on the opposite page, and CIRCLE the location.

1. I left my  and went to a . Then I crossed the street.

Where am I?

2. I left my and skated past the , the , and the . I went around the , then skated

straight to where I was going. Where am I?

3. I left my to drop off my son at . Then I went to the . Then I made one more stop. I did not pass the

or the . Where am I?

Hidden Shapes

FIND each shape hidden in the picture. DRAW a line to connect each shape with its location in the picture.

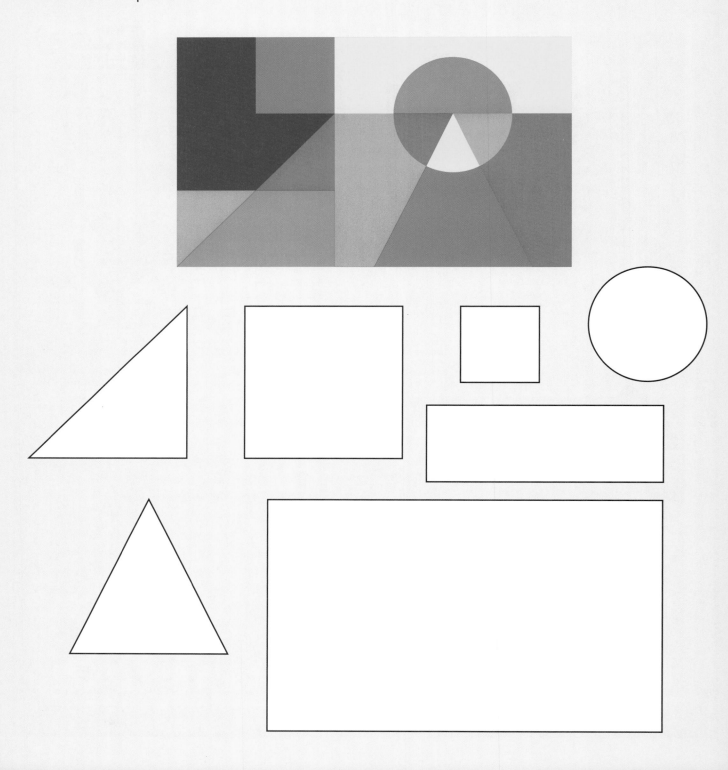

Spiraling Sequence

DRAW and COLOR shapes to finish each pattern. Can you finish the spiral to the center?

Shape Shifters

Use the pattern block pieces from page 117, and PLACE the pieces to completely fill each shape without overlapping any pieces. See if you can solve the puzzles different ways!

Treasure Hunt

There are many pirate treasures, but only one is real. FOLLOW the pirate's directions, and DRAW an X on the correct treasure.

I hid my 🧰 in just the right 🪨. First, I walked from the ⛵ to get fresh water in the 🟦. The 1st 🪨 was too tall. So was the 2nd 🪨. Then I found a 🚩 to climb toward a 🪨. On top of the 🪨, I picked the 🪨 with the 🌴 below it. Can ye find my 🧰?

County Fair

First prize at the Cooper County Fair goes to the longest carrot. CIRCLE the carrot that wins first prize.

Animal Sort

WRITE the numbers 1 through 4 next to each animal, where 1 is the shortest and 4 is the longest.

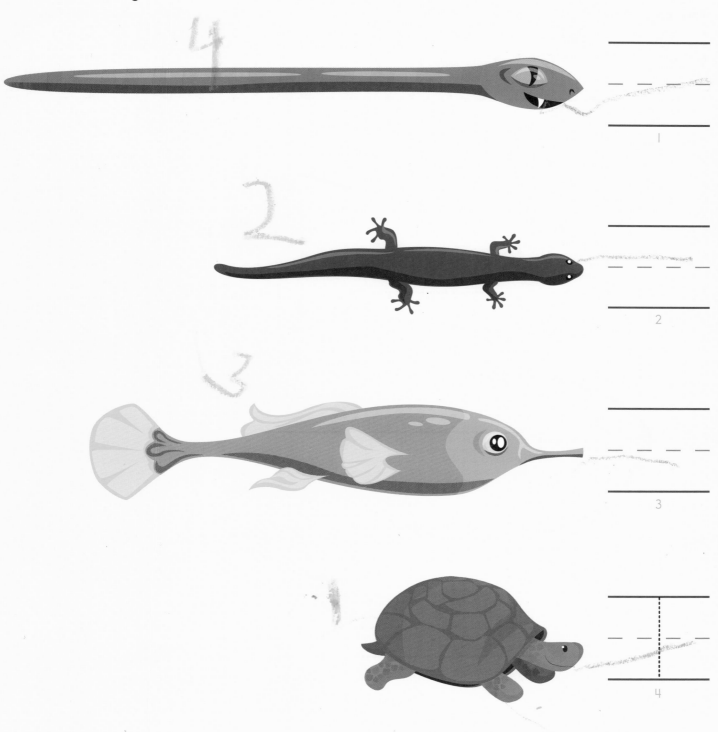

House Hunt

HUNT around your home for six things longer than the crayon. WRITE what you find.

Crayon

_____

- -

_____

- -

_____

- -

_____

- -

_____

- -

_____

- -

_____

House Hunt

HUNT around your home for six things shorter than the pen. WRITE what you find.

County Fair

First prize at the Cooper County Fair goes to the heaviest pumpkin. CIRCLE the pumpkin that wins first prize.

Balancing Act

To stay balanced, the Brothers Brim must hold things that have the same weight. DRAW lines to connect objects that could keep the brothers balanced.

Spoon Flip

BALANCE a spoon on the end of a stack of books by placing a stack of pennies on the end of the handle. PLACE things of different weights in the spoon and see what will flip the spoon and what won't. WRITE what you find.

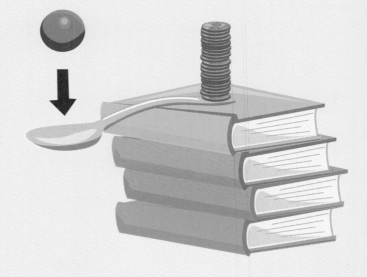

Did Flip

_____

_____

_____

_____

_____

_____

_____

_____

Did Not Flip

_____

_____

_____

_____

_____

_____

_____

_____

Pick for Packing

Help Alison pack for her weekend trip. To keep her suitcase from getting too heavy, she can't bring anything heavier than the big book she's reading. DRAW lines from the objects she can pack to her suitcase.

Build and Compare

CUT OUT each shape on the opposite page. FOLD on the dotted lines, and GLUE the tabs to construct each box. WRITE the answers to the questions.

Look at the two boxes.
Write the color of the one
you think will hold more.

FILL the box that you think holds less with cereal. Then POUR the food from that box into the other. If the food doesn't come to the top of the box, the first box holds less. If the food overflows, the first box holds more.

Which box holds more?
Write its color.

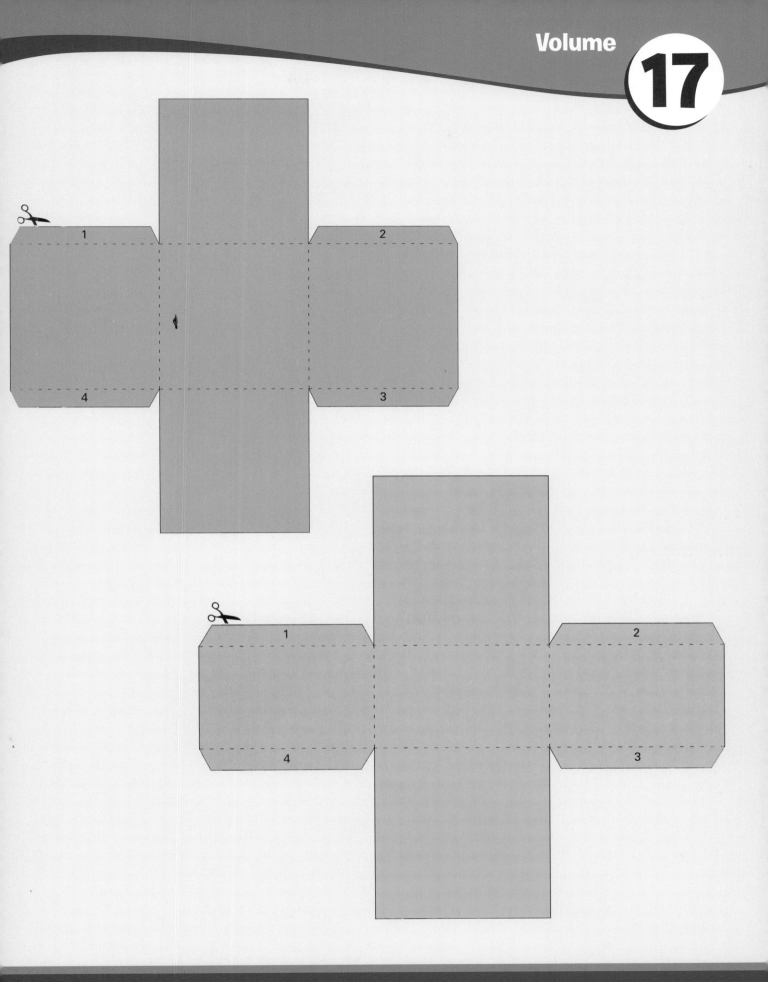

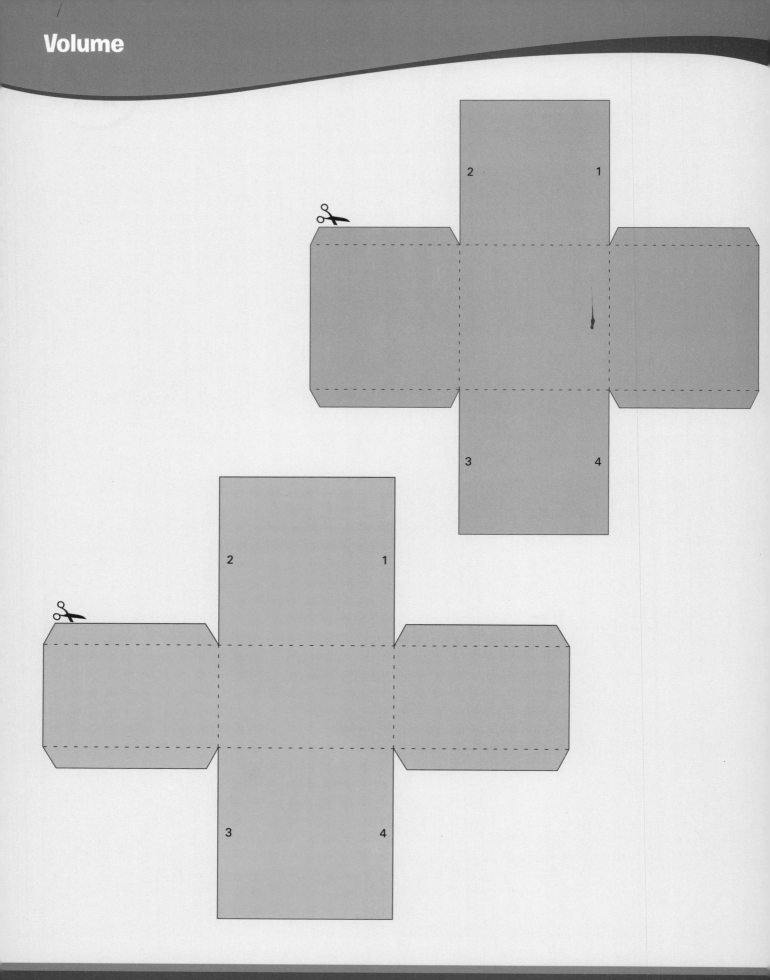

Pick for Packing

Help Ricky pack for his airplane trip. He can only bring things that will fit in his small suitcase. DRAW lines from the objects that will fit to his suitcase.

Shortest to Longest

WRITE the numbers 1 through 6 next to each object, where 1 is the shortest and 6 is the longest.

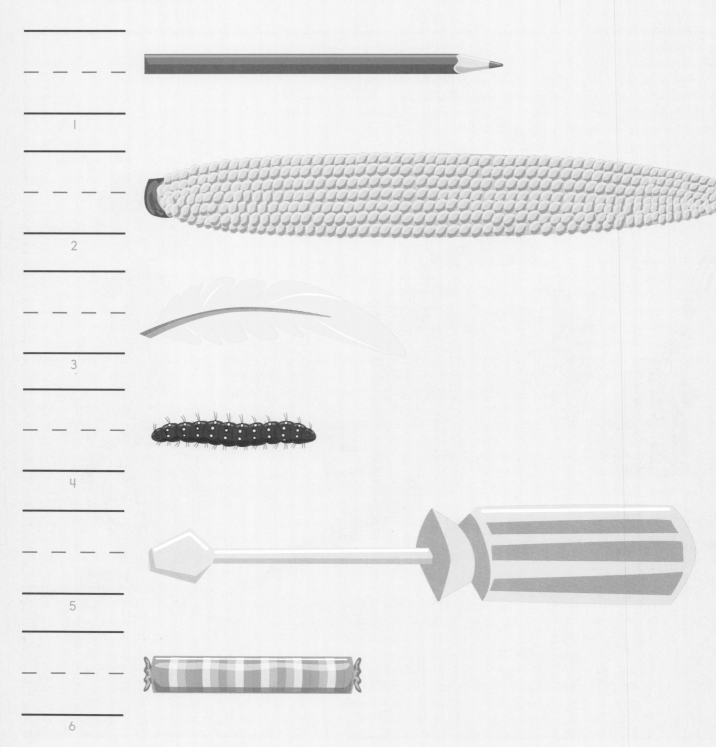

Lightest to Heaviest

WRITE the numbers 1 through 6 next to each object, where 1 is the lightest and 6 is the heaviest.

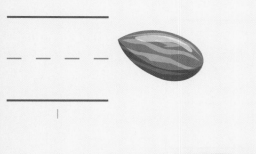

_____
- - - - -
_____
1

_____
- - - - -
_____
2

_____
- - - - -
_____
3

_____
- - - - -
_____
4

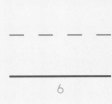

_____
- - - - -
_____
5

_____
- - - - -
_____
6

Smallest to Largest

WRITE the numbers 1 through 6 next to each object, where 1 holds the least and 6 holds the most.

1

2

3

4

5

6

Spinners

CUT OUT the spinner. BEND the outer part of a paper clip so that it points out, and carefully POKE it through the center dot of the spinner. You're ready to spin!

This spinner is for use with page 7, and the reverse side is for use with page 17.

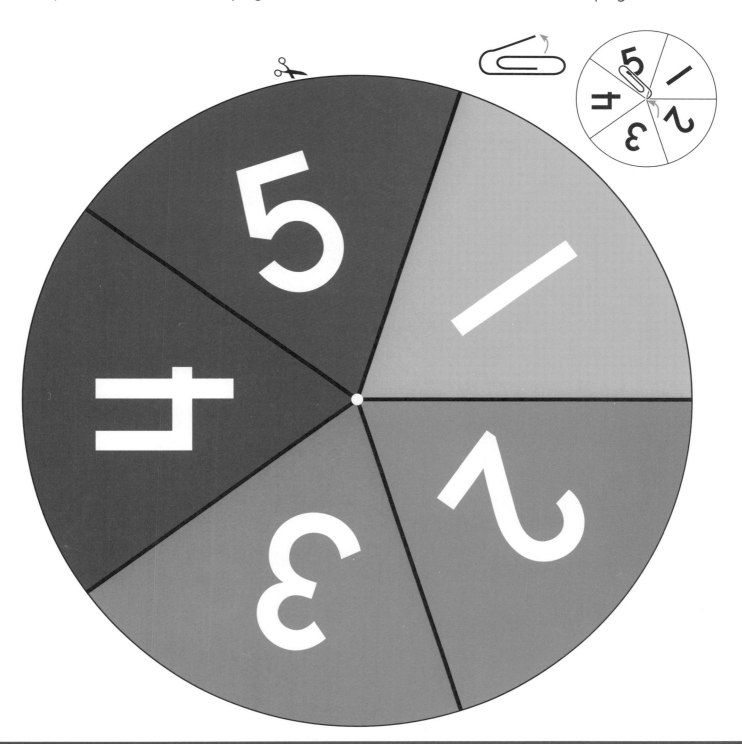

Use the spinner on this side for page 17. Pull out the paper clip from the other side, and poke it through the center dot on this side.

Dominoes

CUT OUT the dominoes.

These dominoes are for use with pages 14, 15, 20, 25, 26, 42, 43, and 60.

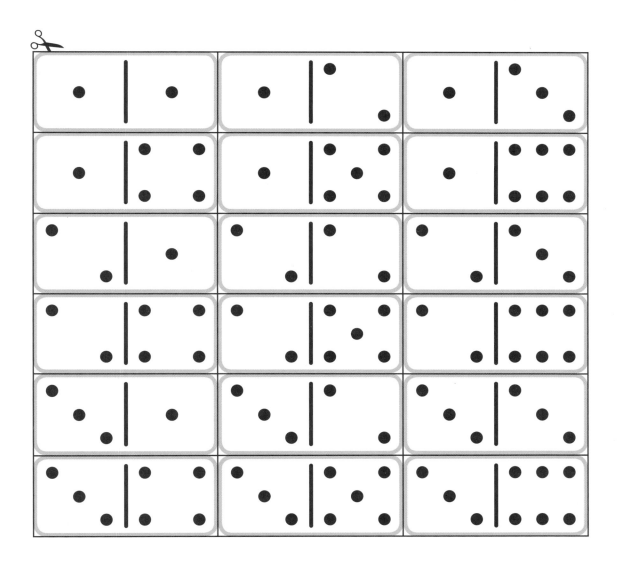

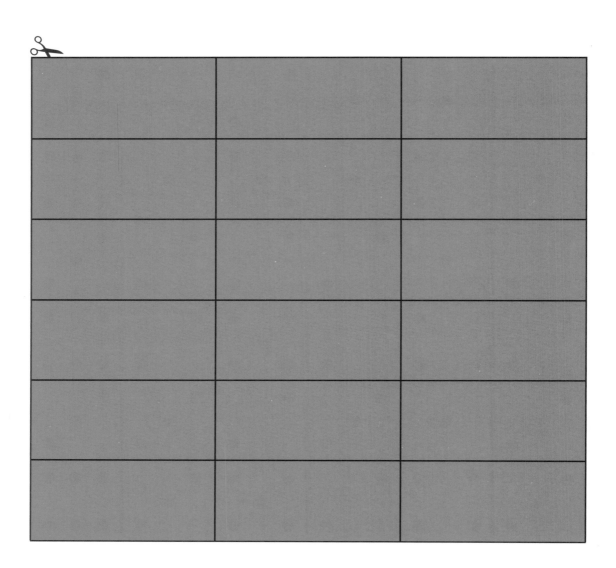

Dominoes

CUT OUT the dominoes.

These dominoes are for use with pages 14, 15, 20, 25, 26, 42, 43, and 60.

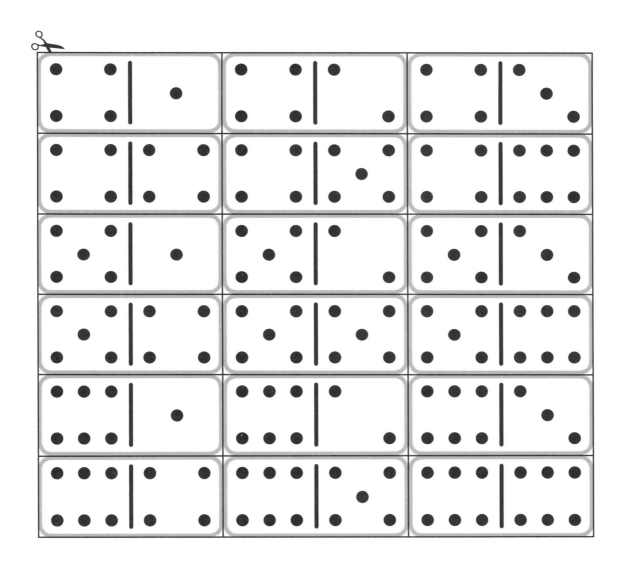

Tangrams

CUT OUT the seven tangram pieces.

These tangram pieces are for use with pages 78 and 79.

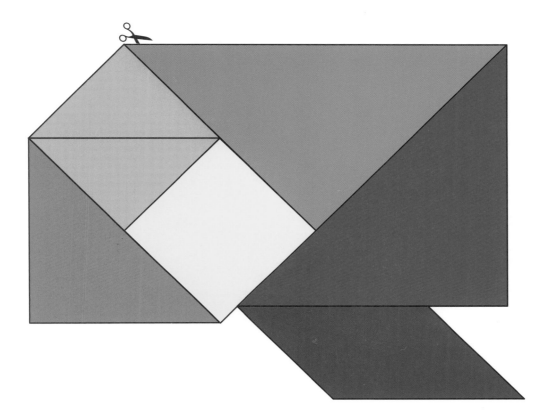

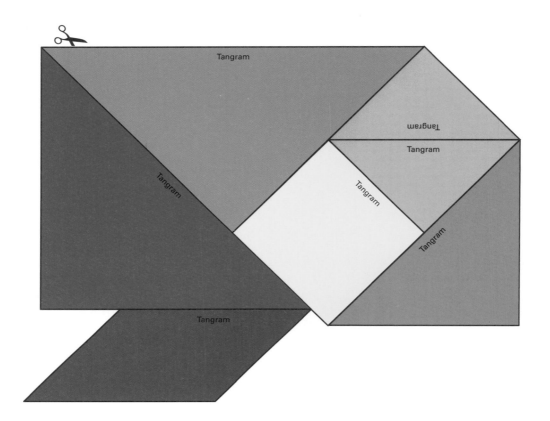

Pattern Blocks

CUT OUT the 31 pattern block pieces.

These pattern block pieces are for use with pages 80, 81, and 92.

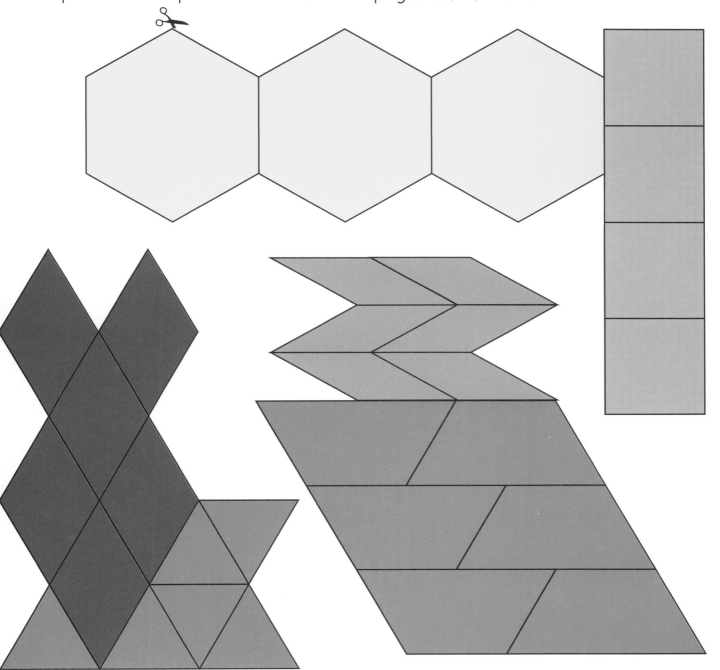

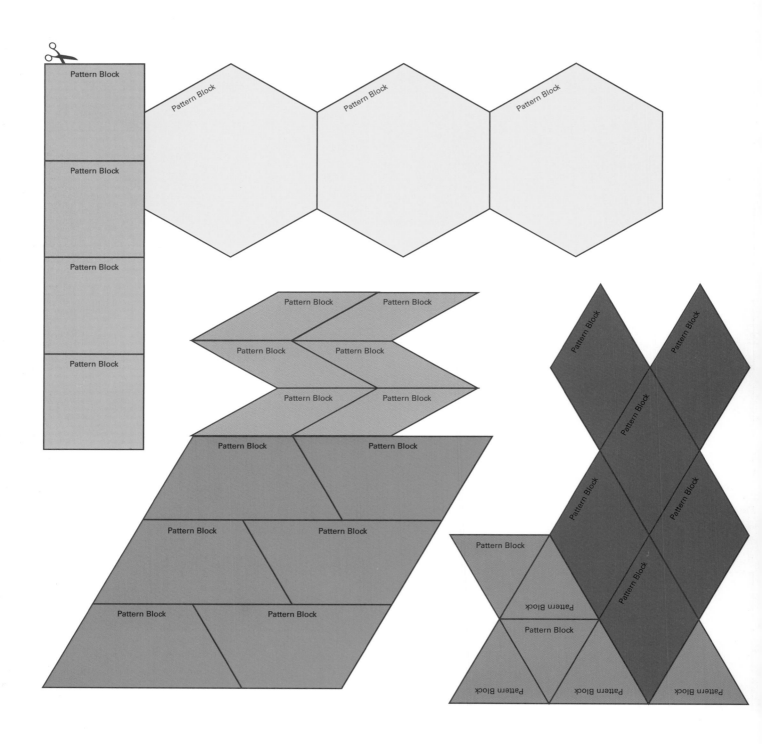

Answers

Page 2

Page 9

Page 16

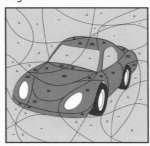

Page 21
Suggestion:

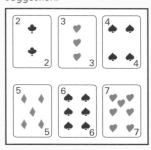

Page 3

Page 10

Page 17
Have someone check
your answers.

Page 18

Page 22

1. 3rd   2. 6th

3. 5th   4. 1st

5. 4th   6. 2nd

Page 23

IF YOU SEE THIS, YOU HAVE
CRACKED MY CODE.

Page 24

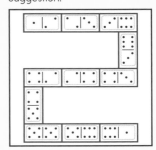

Page 4

Page 11
Have someone check
your answers.

Page 13

1. 4   2. 6

3. 7   4. 8

Page 14
Suggestion:

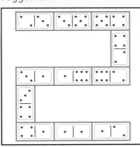

Page 19

Page 5
Have someone check
your answers.

Page 7
Have someone check
your answers.

Page 8

Page 15
Suggestion:

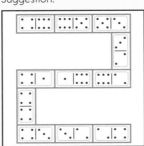

Page 20
Suggestion:

Page 25
Suggestion:

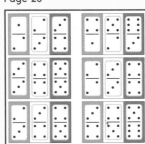

Page 26

119

Answers

Page 27
1. 5th 2. 3rd 3. 1st
4. 7th 5. 2nd 6. 4th
7. 9th 8. 8th 9. 6th

Pages 28-29

Page 30

Page 31
Have someone check
your answers.

Page 33

Page 34
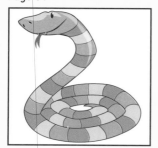

Page 35
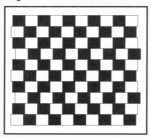

Page 36
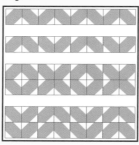

Page 39

Page 40

Page 41
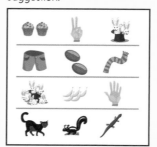

Page 42
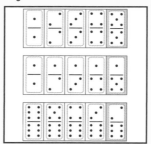

Page 43

Page 44
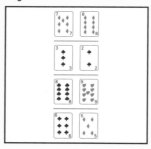

Page 45

Page 46
Suggestion:
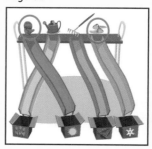

Page 49
Suggestion:
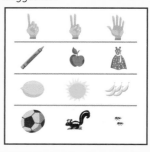

Page 50
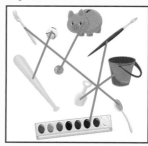

Page 51

Page 52
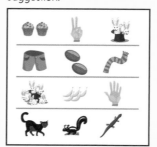

Page 53

Answers

Page 54

Page 57

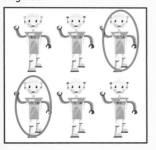

Page 58

Page 59

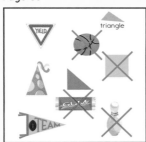

Page 60

Page 61

Page 62

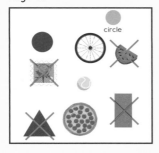

Page 63

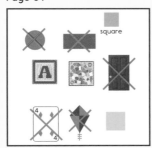

Page 64

Page 65

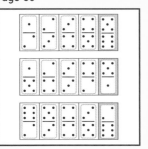

Page 66

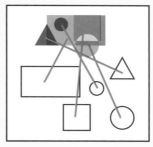

Page 67
Have someone check your answers.

Page 69
1. 4   2. 4   3. 3

Page 70

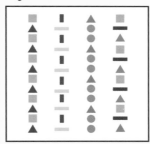

Page 71

Page 72
Have someone check your answers.

Page 73
Have someone check your answers.

Page 74
Have someone check your answers.

Page 75
Have someone check your answers.

Page 76
Have someone check your answers.

Page 77
Have someone check your answers.

Page 78
Suggestion:

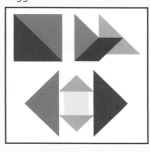

Page 79
Suggestion:

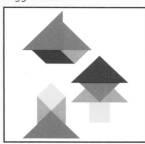

Page 80
Suggestion:

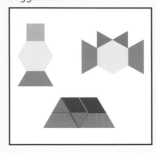

Page 81
Suggestion:

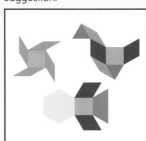

Answers

Pages 82–83

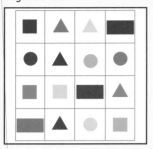

Page 84

Page 85

Have someone check your answers.

Page 86

Page 87

Pages 88–89

1.

2.

3.

Page 90

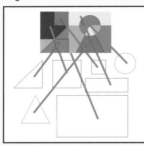

Page 91

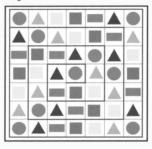

Page 92
Suggestion:

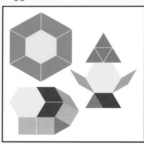

Page 93

Page 94

Page 95

1. 4   2. 2
3. 3   4. 1

Page 96

Have someone check your answers.

Page 97

Have someone check your answers.

Page 98

Page 99

Page 100

Have someone check your answers.

Page 101

Page 102

blue

Page 105

Page 106

1. 4   2. 6
3. 3   4. 1
5. 5   6. 2

Page 107

1. 2   2. 1
3. 6   4. 4
5. 3   6. 5

Page 108

1. 6   2. 1
3. 3   4. 2
5. 4   6. 5

$(x - 17) \times 2 = x$

$2x - 34 = x$

Dau: $2x - x = 34$

$x = 34$

$(34 - 20) \div 2$

CUT ALONG THE DOTTED LINE